Dorkordicky Ponkorhythms

Kwame Okoampa-Ahoofe, Jr.

Dorkordicky Ponkorhythms

Kwame Okoampa-Ahoofe, Jr.

Turn of River Press

Library of Congress Catalog Card Number
97–060862

ISBN 0–938999–09–5

First edition

Turn of River Press
6 Rushmore Circle
Stamford, CT 06905–1029
203–322–5438

To Nana Yaa, Abena, Kwabena, Afua Birago and all their cousins of our third extant generation, who are also our future's bridge and pride, and of course, also, their grandparents, the mediators with our immediate past.

Contents

Preface

In courteously presenting to the world the poems herein, the mouth-watering temptation to over-explicate both the obvious and immutably cryptic has been religiously avoided. My belief has always been that any unsolicited attempt to explain the poetic artifact on the part of the literary artist constitutes a mortal failure; for any creative product demands nothing short of a multicentered appreciation, which is another way of saying that the caliber of "beauty" is co-extensive with the critic's judicial temperament. On this score, therefore, *Dorkordicky Ponkorhythms* is called upon to justify its own bibliographical significance apart from its spiritual and corporeal conduit. Thank You.

Kwame Okoampa-Ahoofe, Jr.
4/28/97

Michelle in A#

In the congestion of my soul,
We lock minds at leisure,
Fantasize, I hope,
About each other,
And tongue-twisted
And tied around
Countless stipples
Of controlled desires,
Confess harmony
In the reserve—

V-shaped book
Is cathartic sheath that engulfs
Thunderous palpitations
Of clam claws
Clasped in dovetail
Slip to perpetually cement
Impending tryst...
The blacker the juice,
The sweeter
This stomach skate

On okra-swept floor
After hailstone rains;
This myth of sartorial nudity
Tickles the mind
To mount reams of papyrus
Only to bounce back
In strip-tease
To headstone—

Resurrection of deeds done
Under sizzling lids
Of chaotic self-consumption;
Shadow after sunset
Collapses into stygian thighs

Of primal night;
Incest,
It was self-rebirth,
When gouged eye-sockets
Signal old-endings
Into new-beginnings—
Mind mashes potatoes on toes
Of wingless quails,
Bats swarm our world,
And the one-eye school-boy
Is our king.

Waco Echo

Echoes of stygian depths
Of bestial descent—
Boomerang echoes
Of foe-meant, poison-tipped assagais
That now sunder our guts...
Waco—
Wacky echo hunt
Of deranged kinsfolk
Led to martyrdom
By a world
That feeds on fire fonts
Of corporeal feasts...
Waco,
Sacrificial smoke beggars Baghdad,
Flippant canine offering
Of rotten eggplants;
Waco echo of a foe's hemlock
Guzzled by kinsfolk...
And our tears
Are desert rains,
For Waco's echo
Is deflected by innocent bones
Of Panama-paving canal carriage
Of pirated pearls;
Cyclical echoes of Cries
Muffled by killer-king's
Stentorian City gong-beater-crier...
And today,
In Georgetown,
They sneer at civilized savagery,
Feral feast-game of cannibals...
Waco echo of Koreshian fumes
Thumb-nose evil
That refinement wrought—
And surely in my heart,
I know Saddam

Does not weep;
And surely in my soul,
I know Noriega
Does not mourn;
And surely in my mind,
I know Castro
Does not chafe;
And surely in my heart,
I know Nkrumah
Does not strain;
And surely in my blood,
I know Qaddafi
Does not weep...
He that has torn
My siblings from the raw womb
Of my mother's hearth,
He shall savor
The sour taste
Of our bereaved...
Waco echo
Of man's lunacy
In shamefaced fumes
Of Koreshian fits—
Waco echo
Of our carnage lust,
Brutal baptism,
Pyro-savagery at noontide,
Wacho echo
Of taxing Texas
Stretches senseless mind taut
To man's futile tact.

4/19/93

Motherhood

for Janet Akosua Edge

Voice is submerged
In gaping womb of earth;
This throat-clearance
Is no soulful song,
Akosua Bodua
It is a dirge—
Flywhisk
Of perpetual pestilence
Of chrome-trapped mind.
Yours was journey aborted
At crossroads,
Anti-climactic journey
Whose outset
Was scabrous endgame—
Earth discriminates not,
She swallows
The perfect
And the blighted;
Earth discriminates not,
She walks the pretty
And the pocked—
Love was seed reaped
From putrid bulbs
Of callous windfalls,
And even then,
To fall,
Fruit must study
Shaft of squall—
Akosua,
Strange
As a mother's Alpine stead
May be,
It still awashes
In very sun-spill

Of toasted Cleopatra;
Akosua,
Let's make love
In groves
Of ancestral battle scars—
Blood bequeathed in birth of death,
It is seminal seed
To tomorrow's deed.

6/16/94

Motherhood II

for Janet Akosua Edge

What shall I proffer you
To blunt
This chthonic bite
Of genetic demise?
One does not
Roast metal money
In a soup-pot,
Else,
I would make bonfire
Of my wallet,
Make tangible to you
The pained recess
Of my soul—
The end just be the start,
Life fed on sweet memories of youth—
Of diapers desecrated without remorse,
For the nightsoil portress
Always devoured
With maternal relish—
Coffin-toting turtle-mom,
Scapegoat of crimes
Committed at twilight of life,
Mother,
You are skull of Golgotha,
Mother,
You inspire wreck
As signal to transcendence;
You will yourself
To festal death of goat-song,
For in the seat of song
Wells redemption—
Immortality is bondage,
Even as mortality
Tickles divinity's crotch—

Akosua Bodua,
You are flywhisk
Of the tailless brute,
Woman,
That part of my soul
That mounted the skies
Encased the main slice
Of my genetic caller's card.

6/17/94

Motherhood III

for Janet Akosua Edge

Silence punctuates grief
And sympathy;
And when our idyllic mind's-eye earth
Aborts labor in clot of spattered maternal blood,
The moorless pain
Kinlessness reaps
Is apocalyptic rebirth
Of primal becoming—
Reflections cooked in midday mirrors
May yet be borrowed images
Of wishful times—
The slaughter,
Nicole,
Was most senseless,
And yet,
In the senselessness of morality
Lies ensconced
Surreality of battered brain-fringe;
Silence
May yet be best balm
To chthonic grief;
Judgments rendered in tawdry straitjackets
Of unexposed hypocrisy—
The sharpest pain
Was also the sweetest balm;
And the mismatches we make
In the drunken name of love,
Like midnight's rat-trap,
Mangles the soul's breath
In the twilight of our faith;
Treachery becomes a fast friend,
And in the sweet thighs of animus
Memory supersedes revenge.

6/23/94

Motherhood IV

for Janet Akosua Edge

She who has not given birth
Can only fathom
Creation's womb
In midnight's clouds;
He who has never fought a battle
Takes sleep for eternity's snore;
She who has never been besieged
By tall tale tellers' tart tongues
Plucks truth from scabrous thighs
Of newspaper leaves—
In moral suicide,
I shore myself against verbal assassins
Of consummated souls;
A neighbor's trial
Is invidious bark of wawa-tree;
It is always another's sole
We pray in crucifix-course;
A villain
Is a hero
Caught plucking pepper
From a pit of dung—
Agony of generous womb
Is repaid by putrid scythed flesh;
Property protects
Even as wealth
Strips the reaper
For a swipe overdealt—
The world is silent with rainbow-banana
Peels;
Fools
Dance to thunderous drum peals
Of battering bats—
As for me,
I have only seen deathly silence

In deafening din;
I have not heard,
I who returned at the crest of cymbal-clash;
I was not there,
I who held the virgin horns
Of the horny ram
In time's mortal tryst.

6/25/94

Happy Birthday, Nkechie

This piebald year of emotions
Humanly mixed,
Of emotions fair and foul,
Of angry words
That are also sails-full of smiles—
This piebald, dead year
Anniversary of our tryst
Beggars inflated goodwills
Of crooked statesmen
On sun's single run—

That we could have come thus far,
It more than signals
How even faster
We could have arrived,
And for good measure
We could still have inched;
The heart that has willed
Entwinement with another soul
Underneath eternity's canopy
Must be coupled
By solemn, unfettered troth
Whose ageless re-grassing
Must be our law of reason,
Edict of our being—
Nkechie,
I am deadly jealous
Who serve my codpiece
In crimson forge
Of your inner-you;
And we must be guarded
Against transient trespassers
Who promise the forever fit
Of bestial passion
On the fleeting crest of lust—
Nkechie,

It is another geoidal encirclement
Against the sun,
And we must be mindful
Of stygian moon-starved nights
That sowed chaos and mistrust
Where unreflective harmony
Should have lain—
This love-trek
Is a tedious argument whose resolve
Must be protean bliss
Engendered by incorporeal beyond—
And yet,
Every debate
Draws us closer
To primal reunion;
Nkechie,
These constant incriminations
And recriminations
May yet
Be pyro-glossal confession
Of displaced renewal
Of sameness
That must dawn
At the pitch-dark hour
Of epiphany
When trivial trysts congeal
In congress to kinship,
And with renewal of our grizzled loins
In fresh foliage,
Arm-in-arm
And earth-to-earth,
In fetal dimple
Of vegetal primal pair,
With pristine birth
Of prepuced time,
Lip-to-lip,
We beget ourselves.

6/14/94

Carleen (née Blake)

My mite
To this forge
Of our future's ladder
Out of this chthonic cavern
In which, against our will, we are held,
Is a shameless replay
Of my rotten
And forgotten past
Subsisting on the gray fringe
Of naked self-invention
Of private parts
Through lucre vaults
Of pressing present needs...

That we shall be rich,
gainsays not the soul,
Nor the scrawny
And still starving body
That must fake mirage aplenty
To crawl out of this stingy sewage
That surveils distended hangover bellies
Of big apple's new york;
Carleen,
One of these days,
We surely
Will have to come up with
Some darn good, marketable plan
To get our tired behinds
Out of this manhole
In the other new york;
Maybe search sternly for our howard's ends
And come up
With crappy purse-cracking muck
To splash ourselves in gold
Even as we dish out slush—
Put some ragged-assed rap group abut,

Even as we scatter our souls
In fecal self-flagellation;
Or,
Maybe
We are some buddhist shrinks
Dumped on this dingy
And dank desert to
Forever lament
Our poverty
In fecund mulch service
Of boesky-vites;
Forever craving,
Never slaking,
Forever griping,
Never gloating,
Forever complaining,
Never harkening to the dark,
Comic beat of titillating forgotten secrets
Which proffered at brazen high noon,
May yet find us
Laughing and guffawing
All the way to the bank
Of common survival,
Preaching the profound prophet's
"Gospel
According to
The growling sad
Stomach's situation."

10/19/93

Carleen II (née Blake)

Invention was your May heyday;
Enthronement of solitude
With smily crown of consortion;
Lifted out of this snowcapped manhole
That holds us hostage
In bottomless ravine of scant—
And like mirage,
When you have nighed to slake scorched throat,
Repairs to further alienation
In endless provocation for endless chase—
The chase,
Carleen,
It began with pelvic affiancement
And bangled, mid-digit...
We cannot be holed with raccoons in ice cups
While at fireside
Felines felicitate on feasts of fowls;
Even the mice play on the royal couch,
While you and I,
Holed up in this dark, solitary cell,
Must sweat
The brawn of our brains
To sate slack missuses whose morning mouth-
washes
Whistle stop our kettle rides—
And we must come up with something,
Carleen,
This deadend drag
Must not forever drag;
And we must come up with something,
Woman,
We also belong
With the terrestrial lot;
And we must come up with something,
Woman,
From whence we come

The brain is also a thinking cap—
And we must come up with something,
Carleen,
The dissolved flesh of our forebears
Also fecundates the earth.

10/19/93

Okru Asante the Fire-Spitter

Okru-Asante the fire-spitter,
You are leopard spawn
From volcanic loins of Adwansaman,
Krobea-Asante,
You are also the son
Of Abuakwa Kuntunkununku,
Kokoo-a-ote-kokoosoo,[1]
Kwaebibirem DehyeE[2]
Akyem-Asante-Kotoko,
You deserve your diadem—
Adusei Peasa DehyeE,
Susubiribi[3]
The tiger-cat,
You fell him on his back
And yet,
He will still land on his feet—
The scabrous paleman from yonderland
Tried to distort your ancestry,
Asante-Denkyira[4]
Adanse-ba[5]
But so mercurial were you
That you bounced right back
Into your own—
Mighty tree who gives shade
To the emasculated and battered,
Osabarima,[6]
If I trip on your toes,
I know I shall land on gold—
Akyem-Asante-Kotoko,[7]
Puny porcupine who is yet
The elephant of elephants,
Aduana-ba Kantinka,[8]
Osuodumgya,[9]
Nana Mawohomereso,[10]
Nana,
Mawohomereso.

5/28/94

[1] Red one who also sits on red; meaning the gold-rich king whose throne is composed of pure gold.
[2] The Duke of the great African forest.
[3] Philosophical Tiger, totem of the Akyem-Abuakwa state of Ghana.
[4] Denkyira is the earliest-known of the modern Akan states of Ghana.
[5] Son of the Adanse state of Ghana, statal radix of the Akyem-Abuakwa state of Ghana.
[6] Champion-warrior.
[7] Kotoko is the porcupine totem of the Asante state and some of its allied Akan neighbors.
[8] Kantinka means brave man or brave warrior.
[9] Fire-quenching water.
[10] Bestir yourself, Chief, demonstrate your authority.

Transition

It is no longer *langue*
But *parole*;
It is no longer
Culture canned,
Frozen and occasionally defrosted
To sate lunatic whims
Of eternal tribalists—
Nke-
Chinyere,
The rustic sylvan bourns are no longer
Definition of race
And difference in group deeds,
Not even saline laps of eternity's
Tongue-rollers ashore;
The city,
Nke,
Is our new bourn,
Cauldron definitiveness,
Inseparable melange;
Arrow of love that defies
Cultivable tongue;
The new order,
Nke,
It is dovetailing of rugged
Intertwinement of ratchet-wheel—
The time turns turvy,
Nke,
And with it change...
We must drink!

10/6/93

Memory

Cube-carapaced tortoise,
We wear our mind-shots in the ruts of our brains,
Nudity that defies mortal rule—
Continuous existence in after-life of remembrance;
What we remember also resurrects the departed;
In your mortality
We abut our eternity;
Wherever you may be resting tonight,
Remember,
There were always those who wished
You could have stayed longer
And mended your mesh-portion
To our dream canopy-net—
You who have poured libation
With your very self,
Let your blood yet wash and drain our tears
In purgation of our mother's womb—
We have reached the abrupt juncture of parting
ways,
And just as sublimity transported us here,
Consciousness may yet train us to ultimate
conjoinment.

6/19/93

Memory II (Tribute to John Mano)

Discourse of minds
Shall form ruts
To signal the path
Of our mortal descent;
Every greeting,
Alas,
Was also
Our parting song—
If lachrymal words
Were sacrifice enough,
Brother,
A billion this much
Will I shed for you.

5/19/94

Besieged

I have no codpiece
For the liquid contents of my life;
Posterity's survival
Rests in the mocking sway of my foes—
I have been besieged by assagaied men
With poison-tipped arrows,
And I must seek refuge at gunpoint
By giving up the sacred innards
Of my cowrie shells—
Trapped by rabied hounds,
I seek confoundment in moral
Self-emasculation;
It rankles
When the midnight becomes
The giver's scourge;
I have deigned, sadly,
To become a spaniel's friend,
And now,
I must risk my lips
Being licked
By a canine's tongue;
The shadows that we cast
Depend on which moment of day
We descend on the battlefield;
Soul of fire
Communes with God;
The gods are silent,
And tyrants become shrines
Unto those who must stay their flesh;
I am toddler grown too late,
Sunshine at midnight,
King who arrived
After the nation
Had already vacated camp;
The fowl has laid thirty eggs,
But Atoapoma,[1]

It is the fetus of earth
In whose sheath
The life of man
Must flow—
I have no codpiece
For the liquid contents
Of my shadow's life,
All because Kwame Atoapoma,
I am a mouse
Among the cats.

2/27/94

[1] Appellation for Kwame, meaning he who recocks and fires his gun unceasingly, or simply, a brave warrior.

Be Grateful

for Nkechie

She gave me a pair of summer socks,
And when I told her
They were too skimpy to weather snowstorms,
She sighed
And gave me
That well-known chilly look and chirped:
Be grateful!
God,
It's a shame
I didn't ask her
What she would do
If she were my oxygen pump—
Then she made *moin moin*[1]
Spiced with alligator-tongued peppers,
And when I sniffed it
And my nose began to drool
And I told her
It was too tough for my throat,
Again,
She gave me that scary well-known look
And roared:
Be grateful!
Golly,
What a shame
I didn't ask her
What she would have done
Had I taken my primal step
In her womb.

2/94

[1] Nigerian bean-meal

Bad Poem

Maybe
It is metered on minor chords;
A bad poem:
It may be
Self-spiting ploy
Of street-soiling goat
And kin-shying deed;
Still,
It needs to be penned,
Bad poem
From good, old poet
That makes the stutterer-novice
Yet hopeful
Of painterly gang—
Bad poem,
Desperate window to thought-starved mind;
Or perhaps
Hopelessly constipated mind
That refuses to give of fertile dung
Because of senseless overfeeding
Of the muse—
And the muse
Must be amused
At this reckless will to
Premature labor
In hopes of birthing
His master's piece—
Bad poem
Like unwanted self-invitation
Of overly-coy hag
Whose orgasmic vintage
Has gone rancid
By automatic recoil—
And yet,
Who said a bad poem
Is no overture

To vintage blues?
And yet who said
A bad poem
Is no refiner-fluid
Of encrusted gem?
And yet who said
A bad poem,
Like an insane's plea,
Is not quintessential
To the gem of freedom's grain?
Who said a bad poem
Is no time capsule
Sown in the futile thighs
Of a tramp?—
Bad poem sang to bad deeds
Of dead-beat dads.
The poem is bad.
And then we know
It is a gist of truth.

5/5/93

Papa Pushkin[1]

Dregs of humanity
That yet plinths
Its solidity—
Purified by enchainment
To sovereign sensibilities;
The scapegoat
Was also beacon
To sublime purgation;
Eternal memory must not be recollected
In tranquility,
That only belongs to the stolid soul
Of drunken speculators;
Charles d'Athenes,
It is in the hemlock
Of mortal dissolution
We find lasting embalmment;
Great-son of Africa's own Hannibal,
Royalty is sole object of our
Committed self-attachment to servitude;
Harbingers of liberty,
We die unto that which we most love,
And in demise,
Become wistful slaves
To reality turned mirage;
Papa Pushkin,
It is not the tongue
That makes the man;
Papa Pushkin,
It is the black-skin soul trashed
And yet also mulched
And composted
To bring life
To white-washed tombs
Of perennial scorn;
Papa Pushkin,
Africanity has no kinship with tongue alone,

Papa Pushkin:
"They want to carve my bust"
Papa Pushkin,
"But I do not want it"
Papa Pushkin,
"My black ugliness"
Papa Pushkin,
"Would become immortal
In all lifeless immobility"
Papa Pushkin,
Such profane self-rejection
Was sublime tribute
To alienation's bestial negation—
And yet,
It was no crass fault of yours,
Papa Pushkin,
Rather,
Incurable benightment
Of a graceless client-folk
Who should have known
The font of mankind's global beginnings
In your negroid loins—
Seminal loins of literate Russia,
Hannibal's quilt-carrier
In this fiery battle of somatic abandon,
Papa Pushkin,
The world must know;
Yes,
Dear direct literary progenitor,
The world must know
In the thick ugliness
Of our songful lips
The world was carted
To reluctant communion
With rectitude,
Papa Pushkin,
Vulgar tongue has no kinship
With corporeal self-cognition,

Papa Pushkin,
The song
That propels man's stolid soul
To the sensible sights
Of Elysian cenacles
Spouts from the ugly
Black bone-marrows of
Besieged and
Shackled Africa
That is truth—
Papa Pushkin...

4/1/93

[1] Inspired by essay from the Anglo-Francophone journal *Afrique Histoire* (Vol. 4, #3,1989), based on the biography of Alexandre Pushkin by Prof. Ibrahima Baba Kake, trans. Wilma Sukapdjo.

Iron-Fisted Mutoko

Mutoko
The iron-fisted
And thistle-thighed,
You tantalized me
With bait of your torments,
Jilted woman,
Licked the stiff corners
Of my bangled dreams
With tears
Of Rubical crossings
You refused to accept...
I have played
With a snot-nosed pup,
Had my lips licked
By anal swathe of swine—
Ebella Mutoko,
Soft, deceptive mirage
Of long, desert trail
That bait-tantalize
Thirsty traveler
To tragic twist—
I am the votive lamb
Whose mane must be singed
As incense-invocation of
Sacred-stool gods;
Woman,
I am the Christ
Whose passionate desire to see
Right topple expedient wrongs
Engendered
Premature harvest of Bobbit-glans—
Ebella Mutoko,
My love
For you
Cannot be reversed,
Ebella,

You tantalize me
With your massive,
Pachydermous
African thighs,
Till confoundment
In clarificatory
Supplication
Is kebab-skewed
In ablution
Of rejuvened love...

4/1/94

Suzuko Morikawa

When my friend
Gariba
Calls
Our Japanese classmate
Suzuki,
Instead of Suzuko
And she angrily
Puts Gariba to right,
I doff
My muted delight
In her courage
To be known
And called
Whatever she wills—
Morikawa,
Like the ring
In "kawa"
Of my Akan-tongue,
Bangled
To invisible heart-strings
Of Suzuko
Of the "Mori"
Corn-dough
Of Ga-kenkey—
Suzuki,
I see her nostrils
Flare with rage,
Her hair
Stand on edge,
And her spear-pointed tongue
Lash
Crazy-horse tongue
Of Gariba's
To dressage—
"Look,
My name is

Suzuko,
Not
Suzuki"—
But, of course
Being
Of Tokyo-born,
What else
Do you expect
A Japan-made
Aficionado
To say,
Confronted with brutal rhyme
At noontide
Sweltering shade
That is a graduate school class?
Suzuko
Morikawa;
Morikawa
Suzuki—
Either way
Automobile memory
Is the same.

4/9/94

Modernity

When old rituals
No longer conjure
Ancestral souls in conclave
Of eternity's static cycles—
Every new birth
Is occasion for strife,
For fatherhood
Is patchwork-quilt
Of multiple matrimony;
Trust
Is cynic kin,
Chastity
Is durance,
As fealty
Is minced meat in doggery;
The crimson fit of passion
Is sole epiphany
Of conjugal sacrament;
The earth tumbles
In multiple temblors,
Lava spills,
And whoever's inner thighs
Are primed for creation
Succumbs to ecstasy—
"I love you,"
Sound...
As long
And hollow
In thrust
As brazen word.

7/26/94

Time to Go Home

It is time to go home
When freedom of expression
Is parceled
And awarded
To the deaf
And dumb—
When those who can speak
And swat
Gadflies
Of vile
Are mauled
By bureaucratic
Sledgehammers
And asked to get
Their broken arms
Together,
Or get themselves
Gathered and
Jettisoned
To make way for those
Willing to sell
Their minds
For a dose
Of self-revulsion
Parading as belonging—
And I must watch my mouth
Or find my wallet
In a garbage bag,
I,
Who have vacated
My royal
Birthcord land
When freedom
Became private prop
For those willed to kill
And oppress by sitting

And flatulating
On the trodden lot;
I,
Who foolishly thought
The world large enough
To make room
For those unwilling
To partake
In this sweepstake orgy
Of salaaming of
The poor
In slobbering saliva drips
Of gratitude—
A deceptive
Lady in a harbor
With vacuous thighs
Beckoning
The battered
With a lighthouse beacon
That only conflagrates and bastes
The conned
Into perpetual
Mortification—
Vilification
Is their culture
Who induct me to refinement
With importunate demands
That I sell
My carcass
For a feast of
Cormorants—
And yet,
How sad
We refused to accept
Their first handshake
Was one
Of our mortal surrender
And perpetual

Denigration—
It is time to go home,
Naked and cold again
Like
Our initial
Awareness—
It is time to go home
When friends plant grenades
In your path
Only to invite you
For the great supper
That must reaffirm
Our inseparable fate.

4/24/93

For "Bryant" (aka Joe)

The mask we wear
Death defiles,
The mask we wear
Death desecrates—
Multiple metacorporeal death
That beggars Christian crucifixion—
If from the Nazarene's crucifix
I must find eternal calm,
Brother,
Yours were best suit to the creed;
Rancid stench of clabbered milk,
Spattered blood
Of gunner's rage,
Silence. Silence. Silence.
Spearpointed silence
That wills eternal woe
On bestial soul,
Woe visited on umpteen
Centuries of incarnadine pedigree
Pledged to cannibalistic predation
In gut-ruled minds;
A death
That makes me question
My Christian life
On libatory script
Of ancestral creed;
And yet,
Early morning
Immolation
To mugger-gods
Must have been signal
To our transient course
In corporeal mirage grove,
Signal to forgotten
Affirmation to seminal radix
Of our beings—

And to be sure,
We are earth,
And this gross greedy deed,
Masturbation of famished desires
That wreaks further craving
For self-negation;
The fratricide
Was most gross
In a land
Where kinship decides
The lawmaker's tort
And the jailer's torque;
Silence. Silence. Silence.
If any grieve
Required response in the redly raw—
The journey was most short,
And the pain
Most unbearable!

5/9/93

Paulette

Hop
Woman
Hop unto the streaming
Truck of my life—
The journey,
It must be wholesome;
Olive-skinned woman,
Flaunt
The taut banners of your balanced behinds;
Caress my arms
With their furry charm,
Woman,
The delicious taste
Of fried onions
Is presaged by its biting
Call in the fresh—
Woman,
You who always call
Me from climactic conference
With my muse,
It is your right,
Pre-eminent woman;
Multicolored ice-cream cone
You lick,
Woman,
I am the ruddy nut
At cornerstone—
Paulette,
Thoughts of touted love for me
Transport my sprite
Beyond stolen trysts
In shaded groves,
At pointed tips of viperous fangs
In passionate
Cancellation of tired romance.

4/23/94

Intimations on Character Assassination

Blackman grown powerful
In a snow-white world,
He shall be killed!
Whiteman grown powerful
In a pitch-black world,
He shall be king!

And I must
Sing my lament
To purgatory's putrid pit,
As my mother
Lull lullabied my light lids
To sweet sound sleep
Of teats flux—
Papa Pushkin
Today I stand aged overnight
Grizzled with gray laughter of death
Scourged and castigated
For calling you my sire—
Blackman grown knowledgeable
In a snow-white world,
He surely must,
He must be killed!
Whiteman grown uneasy
In a pitch-black world,
He surely must,
He must reign supreme!
Black self-knowledge is rude,
It must hang
By its cankered balls;
White self-wash is gold,
It must bangle
The stupefied captive's hand.

I,
Who am dead desperate
For dinner
In the dingy crotch of a tramp,
I surrender
For perpetual damnation
At apocalyptic command
Of those who willed
My slavocratic doom
From Shango's primal forge—
Black self-knowledge is rude,
It must hang
By its cankered balls;
White self-wash is gold,
It must bangle
The stupefied captive's hand.

I,
Who do not deserve
To lick my captor's spittoon,
My eyelashes
Have trampled
My brows,
I,
Who must keep salivating
For mossy crumbs
Of my night rider's bread,
I have polished my nails
In presumptuous gunning
For sumptuous supper
With the cold killers of my kinsfolk—
Blackman grown powerful
In a snow-white world,
He shall be killed!
Whiteman grown powerful
In a pitch-black world,
He shall be king!

12/12/93

I Plant My Seed

Peter,
I plant the chasteful womb
Of my daughter,
Maria,
In the strength
Of your will to fatherhood—
She shall be mother
To all
That God wills
In your cameral forge;
In your marital warm-embrace,
Time's capsule
Shall yet find us alive
And loving
And growing
And maturing
Tomorrow
In the sunrise of our spawns
At sunset
When the mother
Becomes the daughter
Of the daughter,
And the father
Becomes recycled son
Unto his lad—
Maria,
This solemn step
Of soul's entwinement unto soul,
Extends our perishable selves
In cyclical forevers
Which we have been heirs
In the cinerous flash
Of our fleeting flight—
Mortality's love,
Maria,
Must be made strong,

Intense
And sound
To banish regret
From a tongue
That has always felt
The touch of care
In public corridors
Like a hawk,
Always flaunting to the world
The safe catch
On quick-sand search;

Peter,
You must stay strong
Even as
Elegbaic crossroads of sneaky alleys
Confound
The unsuspecting heart
At the peak of bliss—
Remember,
Peter,
She is the one
Who was your own
In the very beginning
Of your womb tryst,
Extended mother-wife
Before the sun
Woke from his stygian sleep
Of genetic pre-being;
Maria,
He is your better-half;
Peter,
She is the one
Who makes you whole…

6/7/94

For Chris Hani

If today
Your departure finds me tongue-tied,
I have also filed
Fatal silence
For Dawn Park's final battle
For atavistic
Re-authentication—

Mention not the name of those
The Cannibals fell,
Let the mind prepare perpetual cannon-balls
To cauterize
Pathological memories
That forever fester
With viral storms;
The hawk may steal broods of
Loud-mouthed, mother-hen;
Silence presages promethean death—
Janusz Walus,
Polarize
Your savagery on the dust-mounds
Of yester-year's
Soil-carriers,
Tragic-thief of civilized souls:
You shall
Wander your restive woe
Till God's eternal demise—
Suffocative silence that summarily severs
Remorseless souls,
You are also the spear
On whose poisoned point
We launch untainted rebirth
Of our clan;
Drunken submission to rapist-gods
Whose sole reward
Is leuco-chaos

In dawn's dark little back—
Coward solicitation of unearned tranquility,
Spirits of my sires,
I bury my shame-shot eyes
In cruel sands
Of naked surrender—
Blood
Spilled from the barrel of butcher's gun,
It spatters veritable gospels
Of tomorrow's sanctioned code—
Janusz Walus,
Polarize my polaroid people
Till splintered to mashed balls,
Tomorrow's seminal liberators
Sprout from the mulch
Of our body's rot.
Silent castigation of a god
Whose godless existence
Finds solemn repose
In suicidal self-sympathy—
We are the children
Of a kinless god,
We are the game
Of Cannibal marauders,
We are the vilified tongue
Of self-word-eating dogs—
We are festal libations for
Wandering royal soil-carrier thieves—
We are the brush
whose total conflagration
Conducts the landowner's toddler tads
To the oblivion
That is our beings—
Africa,
If you must
In erotic slant
Sodomize your foe,
At least

You must know
The sempiternal soul stillness
Of self-willed clasp.

4/12/93

Insinuations

Sorry,
Vicarious murderer
The gods of probity
Have blunted your snaky-spite;
If I must die
Let it be for right and just,
Anger refined to cauterize
Half-millennium, abscess-rot
Of our collective fate;
And yet,
The clouded mind of the enchained
Create self-mockery
To utter disgust
Of even the foe's kinfolk—
If I must die
For brandishing truth,
My soul shall roar with storms
In the sinister shadows
Of boomerang's bang,
For the blood of the pacifist
Must be death-fear
Of the coward-killer;
Soul brother,
If I must die
From the dense sabre strike
Of hatred,
Indeed,
I shall come
As thought-scythe,
Chop your scabrous
Refusal to come into your own
And refine your perpetual guilt
To the pen-point
Of eternal self-rediscovery;
Your gun,
You say,

Has rusted flabby
Like octogenarian penises
That actualize orgasm
In desire's dying sparks;
I shall oil them
With nettle-oil,
Cause your sick soul
To suffer Davidian inferno
In the suicide dance that you,
In your savagery
Will upon my goodwilling pate—
And if I must die,
Why not,
I shall die
The gallant death
Of the healer-prophet
Who makes my carapace casket
With turtle kin—
My soul
Has died
And kissed
Glory
That comes
With ill-quenching sprite;
If I must die,
Surely,
I shall live!

4/23/93

Self-Obit

Casket carapace turtle,
I sing my daily death
At the hands of those
I have willed to free,
Free from the mortal fetters
The dummified brain assumes—
Make me prisoner of death
Brother-of-bad-faith,
For in throes of death
I assumed
Oneness
With eternity—
And I must feed you
With knowledge and truth,
Truth,
Even as you make saw dust
Of my being,
And I must pray
For your health
Even as your fitness
Sublates my life,
And I must weep
For the joy that
Surpasses understanding,
For to understand
Is to know
The painful truth
Of an ingrate-world;
I am the desperate brother
Who will my blood
Unto ablution
In the desolate waste
Of the war's aftermath;
I sing my daily death
Of the mortal snake
That fringes our cloths;

I sing my daily death
Of the far gone dead
Who hang on to life;
Of selfless love
Requited with hate;
I sing my daily death
Of time's severed cord;
I sing my daily death
Of those who must be saved,
Of those who know they must,
But who also know
Repair is beyond recall;
Casket carapace turtle,
The earth
That gave you breath
And light,
It is the same womb
That proffers eternal repose—
My soul is forlorn,
My soul is humbled,
My soul is mortified,
For I have begged life
At the scabrous hand
Of my fellow dust,
My mouth is saw dust
And life
Has become a rotten dream—
I am the rot
That humiliation brings;
Mine is a world of turd—
I am the turd
Which manures frustrated tempers,
I am the turd
Who brazenly dare
To revive the dead,
And yet I'm dead
Who imparts truth of life—
Every tedious argument,

Eliot,
Must end with a period's smudge;
I am the period which
Inarticulate me planted in the
Sultry mid-point
Of a pointed question;
And still,
I must sing my daily death
I,
Who I've seen
The suicidal curse
My overseas cousin has endured;
I must sing my daily death,
For I reap existential cancellation
Wherever I have immolated my heart
In revival rituals
Of collective self-recovery;
Let me aim my fury
At time's treacherous toll,
The ambuscation
Of pedagogical discretion
Until competent evaluation
Plays hide-and-seek
With self-preservation;
I have been raped
Of my godward mind,
I have been shorn of my faith
In the down—
I sing my daily death,
I'm the potter's field.

4/15/93

If Charles Could Be...

If Charles could be tampon
To Camilla,
The world would go to bed
Rest assured of its eternal flux
In the rink of nuclear sleet;
Two-dozen clamactic trysts
Cannot be rugged in a blink
Even if diadem were alchemy
To flip our globe
On endless coital course—
And if Charles
Could truly be tampon
To Camilla,
Gee
Whiz
Lord,
On the day of the moon's
Sanguinary renewal when creation's forge
Anvil has missed its grist,
Such sumptuous feast
Shall he have,
Even if all the world went to hell,
He would still dream
Of inviolate beginnings
In prepuced feral womb—
Let love sing
Its own off-beat song,
For the laurels men bestow
May yet decay to manure
That which by itself becomes...
Love
Stolen in midnight wake,
It is regained
Albeit despoiled
At noontide hour—
Take what the greed of men bestow,

But by all means,
What god has willed,
Let none uncleave.

1/21/94

Scholarly Rat

A poet has cast
His scavenging mesh-finger
Through a jumble mount of scripts
And hauled,
With grime that age spawns,
A dehydrated dead rat;
Gee,
Never knew a rat
Could delve so deep
Our crypt(ic) hieroglyphs
And die musing mangled contents
Of human mind—
The rat,
It reeked
Of fetid forgotten turd
Piled,
One upon another,
As horsing copulation
Of scrambling footballers
At scrimmage-time.

Renewal

Uncle Soyinka,
As for this case
For once
Lionic-bale is wrong:
It is *not*
'The old (that) must flow
Into the new,'
Rather,
It *is* 'the new that must flow
Into the old'...
For the new,
Concealed in the old
Hastens decay into rich
Regenerative transmission of
Bubbling posterity,
For these are the limber waist
Of quintupletic tomorrow,
Uncle Wole,
Tomorrow can only be renewed
Through intelligent corruption
Of today's time capsule;
The jewel must course through
The wobbly veins
Of the grizzled Lion,
And like condom
Before the blowout,
Bear live-witness
To morrow's birth.

2/15/93

Language Arts

If *art* be creation,
Interpretation,
Appreciation and
Criticization,
There is no tongue—
For language is no art
In creation of nature
Which is a defiant force of forgery—
Creation is parody
In failed divinity;
Language...
That which in the beginning
Encoiled god's tongue
To taper our world
Which in creation
Birthed itself
In the cyclical shape
Of insubstantial essence
—Language arts:
All of us in actualized
Pre-formation.

4/11/93

Ogbanje

to Chinua Achebe

Refusal of canopied mushroom's
Return to cradle's dunghill forge;
Black-birth-cord thoughts
Filtered through wobbly tongue
Of a stranger's throat;
They knew how to stanch the race—
Death by assimilation;
Atavistic re-assertion of distorted sensibilities
That must swear sempiternal self-severance
To seed survival—
And yet,
It was not sheer severance
That engendered insanity
But centrifugal celebration of chthonic disconnec-
tion—
Immolation to nihilism
In masochistic plea
For negative salvation;
Ogbanje,
Short-sighted will to transient subsistence
In prelude manure tribute
To wholesale putrefaction—
The return is banishment
Breeding anthropological self-consumption;
And yet
In self-eatery
We extend our spawns—
Obi Okonkwo,
We have stood arsenic drink to each other
Only to make scions of maggots—
And the venomed snake
That will despatch,
It was planted
With the passionate seeds of our beings;

Obi Okonkwo,
We have trigrituded our negritude,
Found ourselves leucotuding
With our eternal mailman.

2/27/93

About the Author

Born in Ghana, Kwame Okoampa-Ahoofe, Jr., who is also the author of *Sororoscopes* (1995), his debut collection of poetry, teaches English at Nassau Community College of the State University of New York. He also regularly reviews books for the *New York Amsterdam News*.